UX Simplified:

Models & Methodologies

Samir Dash

digibooks

2014 India

UX Simplified: Models & Methodologies

By Samir Dash

Visit author's homepage at *samirshomepage.wordpress.com*

Digital Editions rights owned by PatternGraphic, India.
First Edition : 2014

PG2014B07
ISBN-13:978-1500499587
ISBN-10:1500499587

Contents

Design is a funny word. Some people think design means how it looks. But, of course, if you dig deeper, it's really how it works. To design something really well, you have to 'get it! You have to really grok [understand] what it's all about.

—Steve Jobs

Why *UX: Simplified* ?

In software industry , just like any other manufacturing industry the final production of final output step of the whole process of software development is critical as well as expensive. So it became necessary for the different stake holders of the process and the final output to figure out the exact "what" and "how" aspect. To cater these needs new specialties such as "interaction design", "information architecture" emerge. These new disciplines became the part of software development life cycle gradually after 90s and put an impact on production design process in software industry.

Software industry is itself in a speedy evolving process. So the methodologies used in production design also gradually evolved to match the changing needs of the time. For example the competition in the software product market gave rise to the trend of publishing quick and iterative versions to the market and gradually improvement of the product based on the feedback of the users on the versions available in the market. The agile methodology in the software production process helped changing the process related goals for the developers and designers involved. In the designers process concepts like "Lean UX" has emerged to meet the need of the competitive timelines and "quick to production" demands through agile process.

In current book an attempt is made to find answers on how to map different UX and IXD concepts to their "ever changing needs" in software production.

The Diverse Disciplines: The ABCs of UX

It wouldn't be wise to discuss about the process and the involved problems and their solutions, before clarifying the terms we will be using in our discussions. So let's start with the definitions and concepts that are core to our context. Yes, we will start with the ABCs of UX, IA and IXD.

User Experience (UX)

User experience (UX) is a convergence of several disciplines. There is no final fool-proof list of disciplines which come together to form it. But yes, there are many popular explanations by social scientists, industrial designers and information architects which describe it as a combination of a verity of disciplines.

The most popular and accepted compilation of disciplines is shown in Fig 2.a where it is represented as a combination of :

1. Information Architecture (IA)
2. Visual Design
3. Industrial Design
4. Human Factors
5. Interaction Design (IxD)
6. Human – Computer Interaction (HCI)
7. Architecture

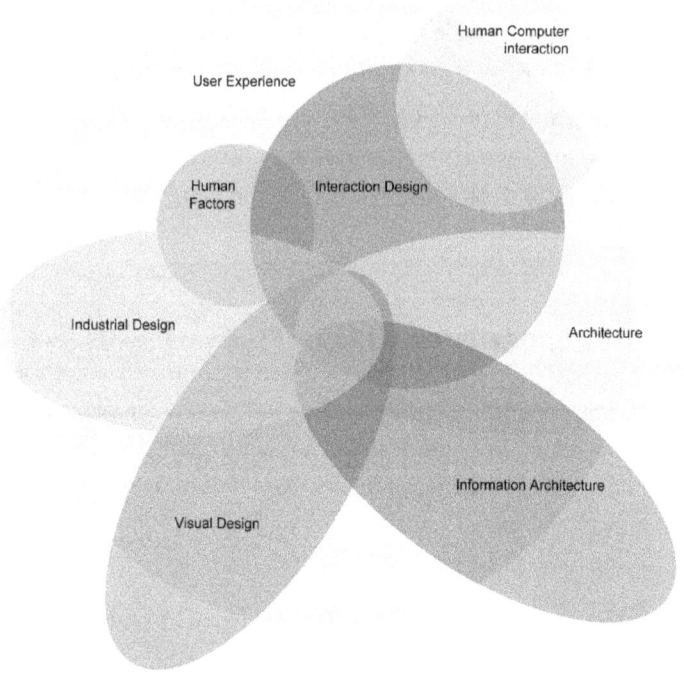

Fig: 2.a

There are variations available to this where commercial aspects are added to it. The following figure represents such a case that takes into account the *Visual design, Branding, Marketing, Customer Service, Information Architecture, Data, Accessibility and Usability* into consideration.

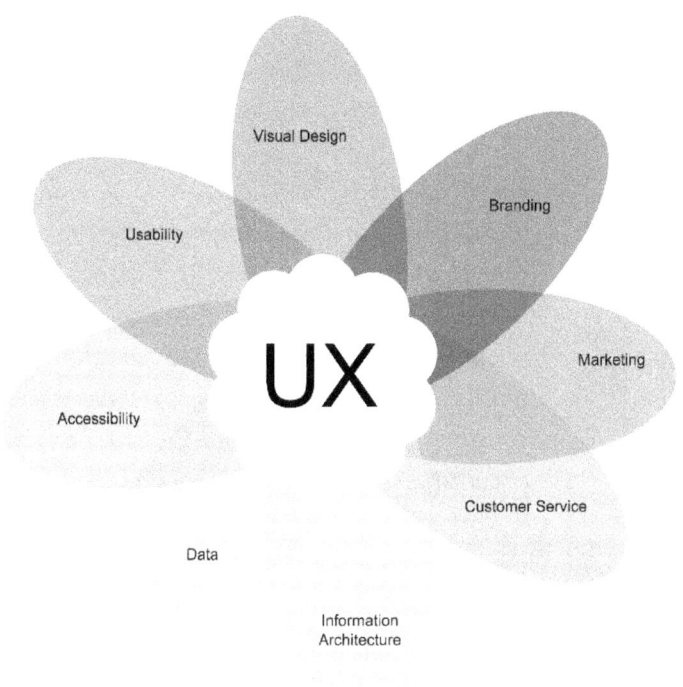

Fig: 2.b

In Fig:2.b , you can see the UX definition has been seen as all those disciplines which can work together to provide a solution that can deliver "customer with a harmonious and consistent experience".

If you notice all the aspects such as "branding" and "customer service" are related with emotional aspect of human behavior. So user experience is also about emotions and psychological dimensions of customer's perceptions about the product. Wikipedia also defines "User Experience" as :

User experience (UX or UE) involves a person's emotions about using a particular product, system or service. User experience highlights the experiential, affective, meaningful and valuable aspects of human-computer interaction and product ownership. Additionally, it includes a person's perceptions of the practical aspects such as utility, ease of use and efficiency of the system. User experience is subjective in nature because it is about individual perception and thought with respect to the system. User experience is dynamic as it is constantly modified over time due to changing circumstances and new innovations."

Similarly ISO 9241-210 defines user experience as:

a person's perceptions and responses that result from the use or anticipated use of a product, system or service". According to the ISO definition user experience includes all the users' emotions, beliefs, preferences, perceptions, physical and psychological responses, behaviors and accomplishments that occur before, during and after use. The ISO also list three factors that influence user experience: system, user and the context of use.

So basically, "User Experience" deals with the "How" factor of the product or system rather than it's "What" factor. Most of the customers buy the product not because simply "what it does", rather the "how" factor takes priority when it comes to choose from a several similar products.

Information Architecture (IA)

While exploring UX in previous section, we saw that Information Architecture (IA) is one of the contributing disciplines which form the total User Experience. During 1960's ,Richard Saul Wurman, an architect by profession having skills of a graphic designer, author and an editor of numerous fine graphics related books, coined the term "Information Architecture". He mostly developed concepts " in the ways in which information about urban environments could be gathered, organized, and presented in meaningful ways to architects, to urban planners, to utility and transport engineers, and especially to people living in or visiting cities". Later these impacted a set of disciplines such as library- and information-science (LIS) , graphic design and in recent years the world wide web (www). During 1990's , with the evolution of the Web, there rose the need to rethink the presentation of library-catalog information as this information has been moved into online public-access catalogs, and in part to the proliferation of information on the Web itself. So during 90's IA has taken on something of a connotation of applying especially to the organization of information on the World-Wide Web.

Basically Information Architecture (IA), all about organizing information in a meaningful way so that the user can easily find it when needed through proper organization, navigation, labeling, and searchingsystems.IA also takes care of the process that ensures that there is no information breakdown or explosion with the scaling of information overtime.

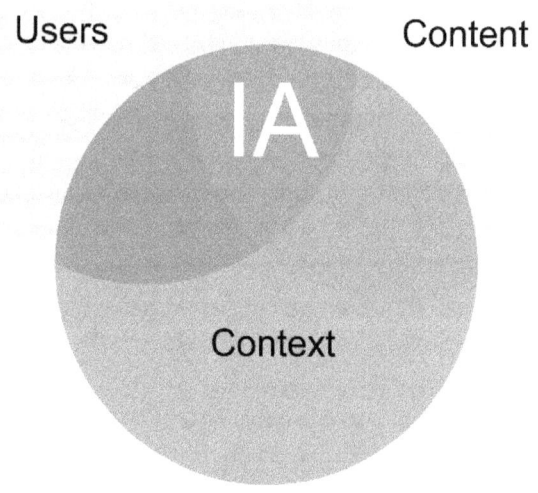

Fig: 2.c

The above figure represents the three basic ingredients of IA, namely:

1. **Users:** This represents those who will use the product or system, their "information seeking behaviour" and their needs. Any of the following roles/skills/features can be applied to them:
    a. Personas
    b. Ethnography
    c. Usability Testing
    d. Expressing User Needs

2. **Content:** This is what is presented to the users through the product or system. This includes the data or information that is offered, along with its aspects such as volume, metadata, structure and organization. Sample skills/roles/features/concepts revolving around content are:
   a. Indexing
   b. Cataloguing
   c. Site Architecture
   d. Writing
   e. Content management
   f. Navigation
   g. Labeling
   h. Search mechanism

3. **Context:** This is what gives meaning to the content that is being served to the user. This can have the attributes like business model, business value, resource, resource constraints, culture etc. Thus the following are features/roles/skills are associated with context:
   a. Defining business requirements
   b. Project management
   c. Business model analysis
   d. ROI calculation
   e. Client management
   f. Technical constraints

Interaction Design (IxD)

Wikipedia defines it as:

> *In design, human–computer interaction, and software development,* interaction design, *often abbreviated IxD, is "about shaping digital things for people's use"*

Interaction design involves attributes of several related disciplines such as :

1. Design research
2. Human–computer interaction
3. Cognitive psychology
4. Human factors and ergonomics
5. Industrial design
6. Architecture
7. User interface design

Interaction design focuses more on human behavior through scientific tools and statistics, even when it is in fact is opposed to the disciplines which focus on "how things are" . Even when it uses tools that span across design and engineering domains, by it's ability to perform "synthesis and imagining things as they might be" makes it a part of "designing" field.

In his book *Designing Interactions*. Gillian Crampton Smith, defined 4 dimensions of Interaction Design, to which Kevin Silver later added the 5th and the most important dimension (i.e. behaviour).

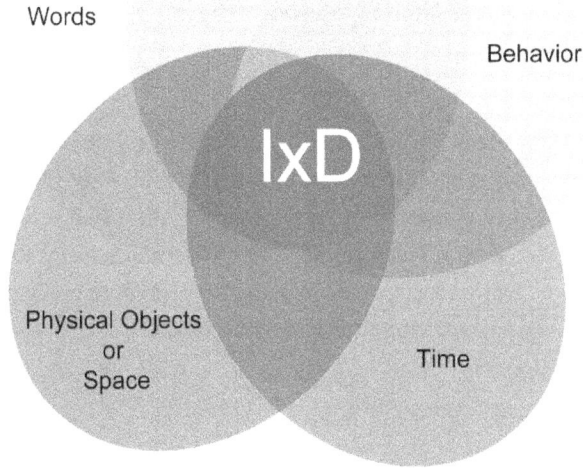

Fig: 2.d

Fig.2d, represents the 5 dimensions of IxD namely:

1. **Words:** This dimension defines the interactions. Words are the interaction that users use to interact with.
2. **Visual Representations:** The visual representations are the things that the user interacts with on the interface. These may include but not limited to "typography, diagrams, icons, and other graphics"
3. **Physical objects or space:** The space with which the user interacts is the third dimension of interaction design. It

defines the space or objects "with which or within which users interact with"

4. **Time:** The time with which the user interacts with the interface. Some examples of this are "content that changes over time such as sound, video, or animation"

5. **Behavior:** The behavior defines the users' actions reaction to the interface and how they respond to it.

Many confuse between Interaction Design with User Interface design, as in most cases interaction design is associated with the designing activities of interfaces. However Interaction Design focuses more "on the aspects of the interface that define and present its behavior over time, with a focus on developing the system to respond to the user's experience and not the other way around".

User Interface Design (UI)

Wikipedia defines it as :

> *User interface design or user interface engineering is the design of computers, appliances, machines, mobile communication devices, software applications, and websites with the focus on the user's experience and interaction. The goal of user interface design is to make the user's interaction as simple and efficient as possible, in terms of accomplishing user goals—what is often called user-centered design. Good user interface design facilitates finishing the task at hand without drawing unnecessary attention to itself. Graphic design may be utilized to support its usability. The design process must balance technical functionality and visual elements (e.g., mental model) to create a system that is not only operational but also usable and adaptable to changing user needs.*

User Interface design is somewhat "the form-giving counterpart to interaction design". User Interface design differs from interaction design primarily in " its focus on the behavior of artifacts rather than the behavior of humans". In UI, the center of all the focus is artifacts that makes the interface, where as in case of IxD it is the human behavior that takes the center stage.

In case of a Software production process the UI is more associated with the Graphical User interface designer. In industrial design case, UI is more tilted towards the industrial designer.

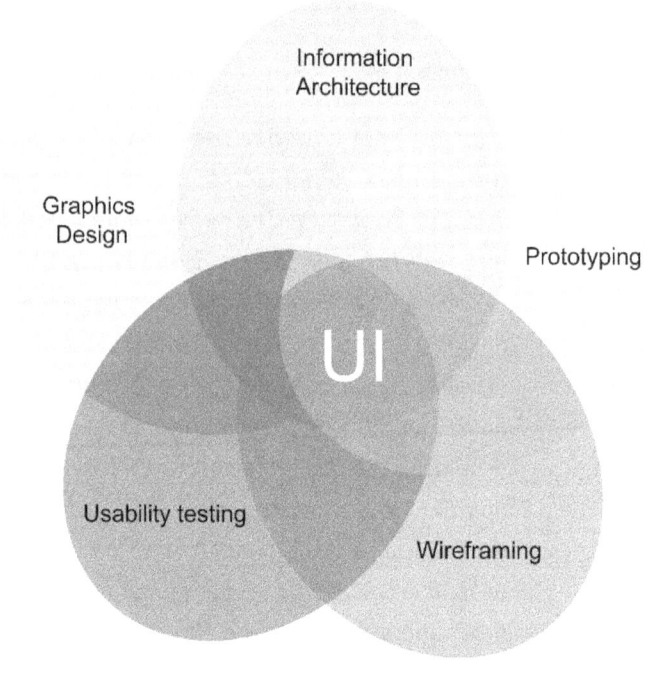

Fig: 2.e

The Fig:2.e, shows various disciplines that contribute to User Interface Design.

Usability and Mental Models: Foundations of UX

What is Usability?

In 1998, the International Standards Organization (ISO),the organization well known for development of standards for industrial processes and product quality, defined "usability" as :

> *the effectiveness, efficiency and satisfaction with which specified users achieve specified goals in a particular environment. (ISO 9241)*

ISO model prescribed the following 3 criteria as components of usability:

1. **Effectiveness:** accuracy and completeness with which specified users can achieve specified goals in a particular environment
2. **Efficiency:** the resources expended in relation to the accuracy and completeness of the goals achieved
3. **Satisfaction:** the comfort and acceptability of the work system to its users and other people affected by its use.

The following figure in the next page represents different standards and attributes.

Usability attributes of different standards or models

1991 Shackel	1992 Schneiderman	1993 Nielsen	1994 Preece et al.	1998 ISO 9241-11	1999 Constantine & Lockwood
Effectiveness (Speed)	Speed of performance	Efficiency of use	Throughput	Efficiency	Efficiency in use
Learnability (Time to Learn)	Time to learn	Learnability (Ease of learning)	Learnability (Ease of learning)		Learnability
Learnability (Retention)	Retention over time	Memorability			Rememberability
Effectiveness (Errors)	Rate of errors by users	Errors/safety	Throughput	Effectiveness	Reliability in use
Attitude	Subjective satisfaction	Satisfaction	Attitude	Satisfaction (Comfort and acceptability)	User satisfaction

One year later , in 1999, **Cosantine and Lockwood** defined 'usability' as:

> *being composed of the learnability, retainability, efficiency of use, and user satisfaction of a product.*

So basically it was an upgrade to the existing "usability" definition, with the addition of two new components:

1. **Learnability:** the product usability can be learned by the user
2. **Retainability:** the product usability an be retained by the user.

Basically the addition of the above two components gave rise to the concepts of "mental model" of the user and it's role in usability.

System Models

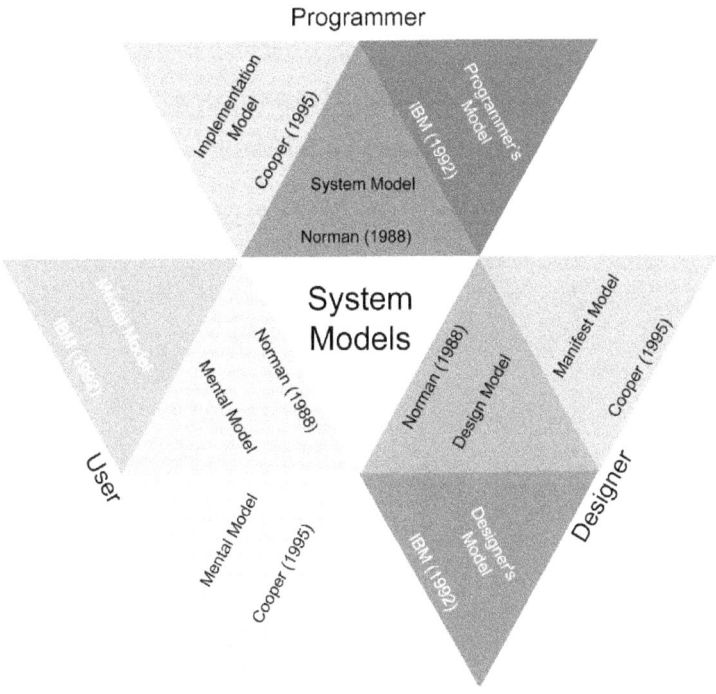

During 1980-90's many "system models" evolved. These models were actually representation of a system from different stake holders' perception, namely: the user, the programmer and the designer.

Among the evolved models, the most notables were that of Norman , Cooperand IBM .

1. **The model based on the programmer's perception:**
    This was the way that a system works from the programmer's perspective
2. **User's Mental Model:**
    The way that the user perceives that the system works.
3. **The model based on the designer's perception:**

The way the designer represents the program to the user, including presentation, interaction, and object relationships.

What is a "Mental Model" exactly?

A mental model is "a person's intuition of how something works based on past experiences, knowledge, or common sense". When you see a book you already know how to use it (i.e. read it) as this understanding of usage of a book is bound with your past experience and the expectation aroused there by using the the book. This typical expectation of how a thing works or the expectation regarding the workflow of an object the user faces is all about a "mental model". When it is the case of an online experience, or a software usage, users expect a certain flow based on both previous experiences, and an expectation on what the experience should be. Understanding and catering to this kind of user's expectation in an intuitive manner is the most critical part of the UX design process.

During "usability testing" of a software product, it is measured against the following five important facets:

1. **Useful:** if the product enables users to solve real problems in an acceptable way and as a practical utility whether it supports the user's own task model.
2. **Findable**: if user can find what he is looking for through his interaction with the system.
3. **Accessible:** if the system can be used by persons with some type of disability such as visual, hearing and psychomotor.
4. **Usable:** if system enables users to solve real problems in an acceptable way
5. **Desirable:** if the user is emotionally motivated to use the system
6. **Meaningful:** It must improve the value and customer satisfaction to be more meaningful in the context.

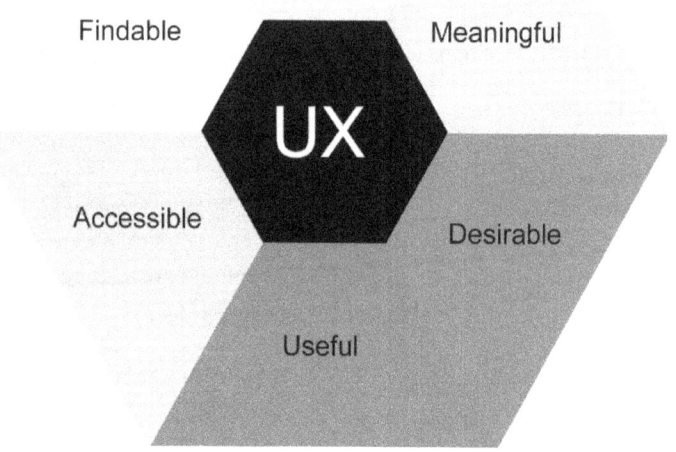

Fig: 3.c

On a close look it is clear that all of these are all related to the users' expectations. If any of these aspects do not match the expectations of the users, the usability of the product/system decorates. So, in UX design process, the most important task is to match or at least bring the design closer to the mental model of the users.

One of the best definition about mental models as they relate to software and usability can be found out in a 1999 article by Davidson, Dove, and Weltz titled *Mental Models and Usability*:

For most cognitive scientists today, a mental model is an internal scale-model representation of an external reality. It is built on-the-fly, from knowledge of prior experience, schema segments, perception, and problem-solving strategies. A mental model contains minimal information. It is unstable and subject to change. It is used to make decisions in novel circumstances. A mental model must be "runnable" and able to provide feedback on the results. Humans must be able to evaluate the results of action or the consequences of a change of state.

Most-likely Mental Model

The problem with matching the interface design and system flow with the mental model of the users is that :

1. There is no fool proof approach that can provide insight into the target users' mental models
2. Different users have different mental models. Different users have different perceptions, past experiences, there by their mental models are most likely not the same.

In real life, an interface will never match up with every mental model because the number of possible models ranges from one to millions. So the trick is to create interfaces to match the most-likely mental models for the target users.

To determine what can be the criteria of a most likely mental model, typically different user personas, research, prototyping and user testing tools and methodologies are used.

Conceptual Model

"Conceptual Model" is a term used to represent the engineered interface that is provided to the user. For example, we can think about *iBooks* app on an iPad as a conceptual model being offered to the user (See Fig:3.b).

Fig: 3.d– *Left image shows the iBooks app and the right image shows the real life physical book.*

To make UX successful, the "conceptual model" is designed to come close to the "mental model".

If the user has read/seen any physical book, then in this case it will be easy for him to use *iBooks* app as it's interaction approach is similar to a real life book where the user can turn pages to read. However iBooks has been designed by some engineer that presents a similar experience to the real book . This conceptual model in this case matches with the expectations of the user who has never interacted with the app,but has some pre-conceptions regarding it .His mental model about the app is formed from his past experience of interacting with the real physical book.

Challenges in Usability Measurement and Metrics

All the models and facets of usability described above have some limitations as it is not straight forward to implement them to some kind of metrics by which the usability goals can be measured. This is mostly because, there is comparatively little information about exactly how to select a set of usability factors to form a metrics to measure in the context of a software development lifecycle having aspects such as business needs and goals, management objectives, resource limitation on product development.

Even though in generic terms "usability" refers to a set of multiple concepts, such as execution time, performance, user satisfaction and ease of learning ("learnability"), taken together, it is still not been defined homogeneously to a level useful for creating a fool-proof metrics.

> *A challenge with definitions of usability is that it is very difficult to specify what its characteristics and its attributes should be, in particular because the nature of the characteristics and required attributes depend on the context in which the product is used. (Alain Abran et. al. , 2002)*

For such reasons there is always a scope to create a consolidated usability model and its factors that can work for creating a metrics useful for software development life cycles.

A List of Factors for Generic and Consolidated Usability Model

As discussed above, here is a very generic consolidated usability model that can be used to create a metrics for a practical usability review.

The suggested model covers most of the commonly reviewed "factors" of different software products and systems which can be customized depending on the context or the need of the project:

1. **Effectiveness:**This factor can be used to measure if the user is able to complete the tasks on product or the system (e.g. a website).
2. **Efficiency:**It measures if the user is able to carry out the tasks, accurately and quickly.
3. Findable: if user can find what he is looking for through his interaction with the system.

4. **Expectations**: this measures if the user mental model matches with the conceptual model offered through the system.
5. **Emotions**: This measures how the user feels while and after using the system .
6. **Satisfaction/ Experience:**This measures if the overall system usage for the user is positive and if the user would like to revisit/reuse the system in case of the need (or will he look for alternatives).This is basically the subjective responses from users about their feelings when using the system.
7. **Productive:** This measures if the amount of useful output that is resulted from user interaction with the system.
8. **Learnability:** This measureshow easy it for the user to master the usage of the functionalities.

9. **Safety:** It measure the level of safety of the user and his information during and after the period of operation
10. **Accessibility:** It measures capability of the system to be used by persons with some type of disability such as visual, hearing and psychomotor.
11. **Usefulness:** It measure is the product enables users to solve real problems in an acceptable way and as a practical utility whether it supports the user's own task model.
12. **Universality:** It measures if the system has universal appeal and enables the users from diverse cultural backgrounds and locale.
13. **Trustfulness:** This specially measures if the user trusts the system for critical usage (such as using credit cards on an ecommerce site)
14. **Meaningfulness:** It must improve the value and customer satisfaction to be more meaningful in the context..

Based on the above factors, usability metrics can be prepared to conduct usability testing on the system. We will discuss about further in the forthcoming chapters. Now let's explore what are the standard processes and methodologies used in a User Experience design in context with standard software development life cycle.

For usability models, typical heuristics methods are applied to measure usability of the system in context.

Heuristics: Measuring Usability

Heuristics Evaluation is defined by Wikipedia as :

> *A heuristic evaluation is a usability inspection method for computer software that helps to identify usability problems in the user interface (UI) design*

The main goal of this is to identify any problems associated with the design of user interfaces. Heuristic evaluations are "one of the most informal methods of usability inspection in the field of human-computer interaction". These are called "heuristics" as these are more in the nature of rules of thumb than specific usability guidelines.

Heuristics usability evaluation method was developed by Jakob Nielsen, who also founded the "discount usability engineering" movement for "fast and cheap improvements of user interfaces" and several other usability methods. There are other usability inventor's who have also prescribed guidelines or a different sets of "heuristics", how ever Neilsen's set are more popular in practice.

Nielsen's methods are defined in his book *Usability Engineering* as follows:

1. **Visibility of system status:**The system should always keep users informed about what is going on, through appropriate feedback within reasonable time.
2. **Match between system and the real world**: The system should speak the user's language, with words, phrases and concepts familiar to the user, rather than system-oriented terms. Follow real-world conventions, making information appear in a natural and logical order.

3. **User control and freedom**:Users often choose system functions by mistake and will need a clearly marked "emergency exit" to leave the unwanted state without having to go through an extended dialogue. Support undo and redo.

4. **Consistency and standards**: Users should not have to wonder whether different words, situations, or actions mean the same thing. Follow platform conventions.

5. **Error prevention**: Even better than good error messages is a careful design which prevents a problem from occurring in the first place. Either eliminate error-prone conditions or check for them and present users with a confirmation option before they commit to the action.

6. **Recognition rather than recall**: Minimize the user's memory load by making objects, actions, and options visible. The user should not have to remember information from one part of the dialogue to another. Instructions for use of the system should be visible or easily retrievable whenever appropriate.

7. **Flexibility and efficiency of use**: Accelerators—unseen by the novice user—may often speed up the interaction for the expert user such that the system can cater to both inexperienced and experienced users. Allow users to tailor frequent actions.

8. **Aesthetic and minimalist design**: Dialogues should not contain information which is irrelevant or rarely needed. Every extra unit of information in a dialogue competes with the relevant units of information and diminishes their relative visibility.

9. **Help users recognize, diagnose, and recover from errors**:Error messages should be expressed in plain language (no codes), precisely indicate the problem, and constructively suggest a solution.

10. **Help and documentation**: Even though it is better if the system can be used without documentation, it may be

necessary to provide help and documentation. Any such information should be easy to search, focused on the user's task, list concrete steps to be carried out, and not be too large.

Engineering and Design Processes:

Usability and User Centric Approach

Usability Engineering

Usability Engineering began to emerge as a distinct set of "professional practice" in the mid- to late 1980s. The majority of the professionals of this practices were from varied backgrounds such as Computer Science or in a sub-field of Psychology such as Perception, Cognition or Human Factors. Today this field is being populated from some newer discipline such as Cognitive Science and Human-Computer Interaction.

Usability engineering, is defined byPreece as

> *'an approach to system design in which levels of usability are specified and defined quantitatively in advance, and the system is engineered towards these measures, which are known as metrics.'*

The whole concept of Usability Engineering focuses on the "metrics for measuring usability".

As the emphasis on usability metrics through "analysis and evaluation"is mostly the soul focus of this process, there is not enough focus on the actual design process. In this process the usability is tried to be attained through "engineering and quantifiable methods and techniques" rather than "designing the way to usability".

Also the "usability engineering"focuses only on providing range of techniques to analyze users, specify usability goals, evaluate designs, but it does not address the whole development process.It has more of a focus on "assessing and making recommendations to improve usability than it does on design, though Usability Engineers may still engage in design to some extent, particularly design of wire-frames or other prototypes".

The usability engineering mostly seen as a separate activity that can be plugged into different SDLC models as a separate set of activities from a process-oriented perspective.

The Usability Engineering conducts evaluations through the following tools and methodologies:

1. usability testing
2. interviews
3. focus groups
4. questionnaires/surveys
5. cognitive walkthroughs
6. heuristic evaluations
7. RITE method
8. cognitive task analysis
9. contextual inquiry
10. Think aloud protocol

User-Centered Systems Design (UCSD)

User-Centered Systems Design (UCDS) is set of "usability design" process focusing on usability throughout "the entire developmentprocess and further throughout the system life cycle". It is based on the following key principle:

1. **User focus**: The goals of the activity, the work domain or context of use, the users' goals, tasksand needs should control the development.
2. **Active user involvement**: Representative users should actively participate, early and continuously throughout the entire development process and throughout the system life cycle.
3. **Evolutionary systems development**: The systems development should be both iterative and incremental.
4. **Simple design representations**: The design must be represented in such ways that it can be easily understood by users and all other stakeholders.
5. **Prototyping**: Early and continuously, prototypes should be used to visualize and evaluate ideas and design solutions in cooperation with the users.
6. **Evaluate use in context**: Baseline usability goals and design criteria should control the development.
7. **Explicit and conscious design activities**: The development process should contain dedicated design activities.
8. **A professional attitude**: The development process should be conducted by effective multidisciplinary teams.
9. **Usability champion**: Usability experts should be involved from the start of project to the very end.
10. **Holistic design**: All aspects that influence the future use situation should be developed in parallel.

11. **Process customization**: TheUCSDprocessmust be specified, adapted and implemented locally ineach organization. Usability cannot be achieved without a user-centered process. There is, however,no one-size-fits-all process.
12. **A user-centered attitude must be established**: UCSD requires a user-centered attitude throughout theproject team, the development organization and the client organization.

The typical process flow of UCSD can be visualized as the following steps (based on ISO/TR 18529:2000):

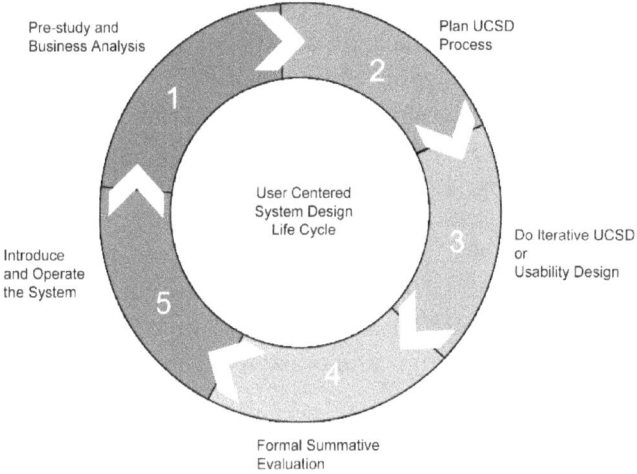

Fig: 4.a

1. **Pre-study and business analysis**: It can be anything from a comprehensive analysis of work procedures, business processes, etc., to a brief statement or vision.

2. **Planning the user-centered systems design process:** includes setting up the project with resources, activities, roles, methods, etc.

3. **Do iterative UCSD /Usability DesignActivities**: The usability design process approximately.

4. **Formal Summative Evaluation**: It covers the usability of the resulting system, as opposed to the formative evaluations used in the usability design process to learn about details in the design .

5. **Introduce and Operate the System**: includes installation, change management, user training, evaluating long-term effects and so forth.

The focus of UCDS is all about "changing the attitude among all professionals involved in the software development process" and these set of 10 principles are key for the "user-centered systems design process" which helps in giving "equal weight to interaction design, analysis and evaluation, combining interaction design, and usability engineering".

Usability Design

The Usability Design is roughly a subset of the UCSD process that matches the "Do Iterative UCSD" step of the UCSD process.

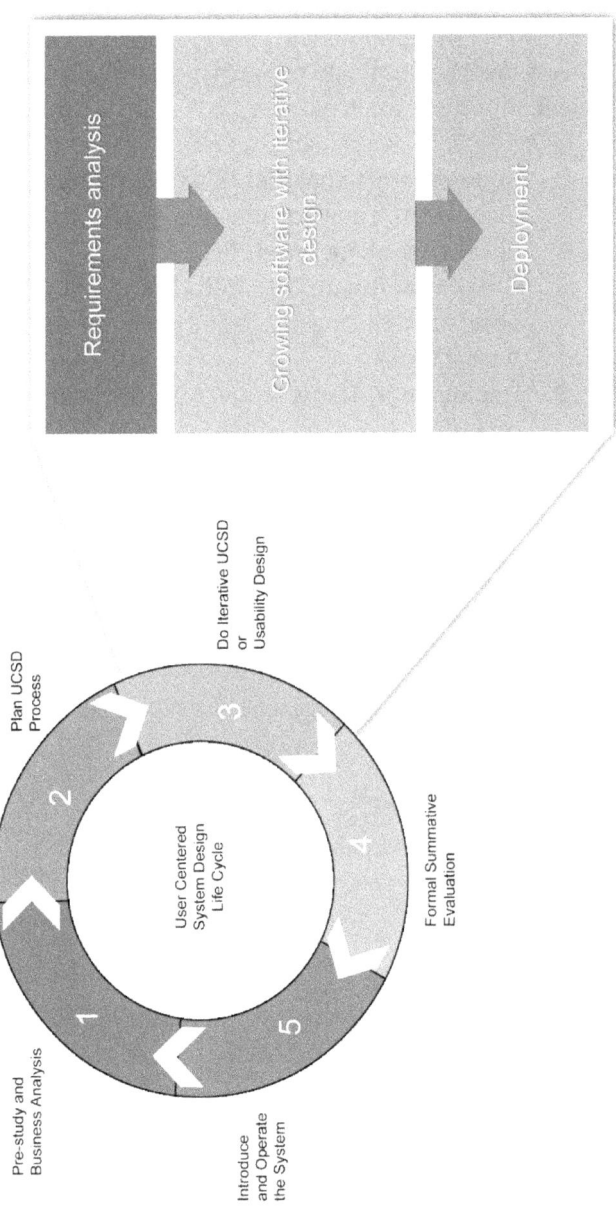

The usability design outlines the steps in the development process involving usability design aspects. The process can be divided into three main phases:

1. **Requirements analysis**: This step is synonymous with planning and analysis phase of typical software development life cycle(SDLC).
2. **Growing software with iterative design**: This is the design and testing phase and development phase of typical SDLC.
3. **Deployment**: This is same as deployment phase of typical SDLC.

User-Centered Design (UCD)

User-centered design (UCD) is a set of design processes in which "the needs, wants, and limitations of end users of a product are given extensive attention at each stage". It is characterized as a multi-stage problem solving process involving designers who take the lead responsibility in foreseeing and solving the usability problems the users are likely to face while interacting with or using the system/product. UCD focuses on understanding the behavioral aspect of the user interacting for the first time so that the user's learning curve in using the system can be evaluated in order to optimize and reduce it. User-centered design philosophy emphasizes on optimizing the product around "how users can, want, or need to use the product, rather than forcing the users to change their behavior to accommodate the product".

Constantine and Lockwood define UCD as :

> '. . . loose collection of human-factors techniques united under a philosophy of understanding users and involving them in design'. . . 'Although helpful, none of these techniques can replace good design. User studies can easily confuse what users want with what they truly need. Rapid iterative prototyping can often be a sloppy substitute for thoughtful and systematic design. Most importantly, usability testing is a relatively inefficient way to find problems you can avoid through proper design'. ('. . . loose collection of human-factors techniques united under a philosophy of understanding users and involving them in design'. . . 'Although helpful,

*none of these techniques can replace good design. User
studies can easily confuse what users want with what
they truly need. Rapid iterative prototyping can often be
a sloppy substitute for thoughtful and systematic
design. Most importantly, usability testing is a relatively
inefficient way to find problems you can avoid through
proper design'.*

Putting it straightforward UCD is all about 4 factors which are
mostly related to the end user :

1. Needs of users
2. Limitations of users
3. Preferences of users
4. Business objectives of the product.

This helps in achieving the following benefits:

1. User satisfaction through more user friendly product
   experience
2. Increase in customer /user loyalty.
3. Making the product more relevant and valuable for the
   user
4. Product / system is more value added as users

Don't get Confused: UCD vs UCSD

UCD is differs from the UCSD in the following areas:

1. **Goal**: The goal of UCSD is more on the process than the user so as to make the final product/system more usable. UCD rather focuses more on "users" of the product and not the design process. More focus is spent on understanding the user and their need.

2. **Process vs. Techniques Set:** UCSD is about system development where as UCD is mostly a set of tecniques and process sets to be used with in UCSD

3. **Perception**: The DNA of UCSD is about changing the mindset of the professionals in the development process so that the designing aspect of usability can be put into practice freely and with higher priority. The UCD process is not about the changing perception about the priority of the design in the whole process.

4. **Broadness**: UCSD covers the whole process that includes the areas which are even not part of "designing" whereas UCD can be seen as a subset of UCSD focusing of the "design process sets".

UCD Models and Process

There 3 different models that support UCD in varying degrees and follow the ISO standard Human-Centred Design for interactive systems:

1. **Cooperative Design**: This involves designers and users on an equal footing.
2. **Participatory Design (PD):** Inspired by Cooperative Design, focusing on the participation of users
3. **Contextual Design**: "Customer-Centered Design" in the actual context, including some ideas from Participatory Design.

All these UCD models involve more or less a set of activities grouped into the following steps mentioned below:

1. Planning: in this stage the UCD process is planned and if needed customized. It involves understanding the business needs and setting up the goals and objectives of the UX activities. Also forming the right team for the UX needs and if needed hiring specialties fall into this step.
2. User data collection and analysis: This step involves data collection through different applicable methodologies such as user interviews, developing personas, conducting scenarios , user-cases and user stories analysis, setting up measurable goals.
3. Designing and Prototyping : This involves activities like card sorting, conducting IA, wire framing and developing prototyping.

4. Content writing: this involves content refinement and writing for web and similar activities.
5. Usability testing: This involves is a set of activities of conducting tests and heuristic evaluations and reporting to allow refinement to the product. However Usability Testing can have its set of steps involving similar activities such as planning , Team forming, testing , review and data analysis and reporting.

All these are similar to most of the steps that fall under Usability Design as UCD can be seen as a subset of process with in Usability Design.

So many processes: What is where?

After going through multi relation models in all these processes and sub process discussed in this chapter, it might be little confusing to visual all the overlapping and dependable process sets. So here is a simple representation diagram that roughly shows the overlapping relations:

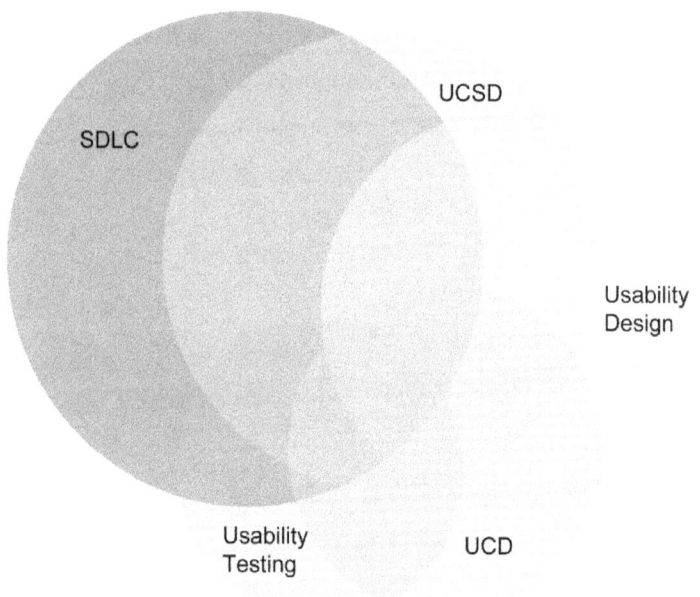

Fig: 4.c

Software Development Life Cycle (SDLC):
Where and How User Experience Models fit in?

Looking at the basic definitions in the previous chapter, one can easily assume that all these concepts are interrelated and there by supplement or complement each other depending on the context. We will explore how these different concepts play a role in a software production process. Also exactly how these components fit into the Software Development Life Cycle (SDLC).

Software Development Life Cycle (SDLC)

To understand what are the processes involved to create experience for a software product, we must first understand is SDLC. There are many diversified methodologies, techniques which are being used different organizations and firms for software development.

However, all these different models have following basicphases in common in their process:

1. **Initiation**: This begins when a there is a need for the software or system to be developed.
2. **Ideation/Conception**: This phase is responsible for the definition of scope of the product. Also all the cost benefit analysis, Risk analysis and feasibility study happen in this phase.
3. **Planning**: A project management plan is formed in this phase. All the plan on how to gather requirements and execution plans at this stage.
4. **Requirement Gathering and Analysis**: User needs and requirements are gathered and analyzed.

5. **Design:** System design is created based on the analysis report of the requirements gathered in the previous phase.
6. **Development**: All the design is converted into actual system or the product. Actual system coding, unit testing, refinement of the program happens in this phase.
7. **Integration and Testing**: In this phase, the system developed is tested against some test- plans to see if the system confirms to the requirements. Bug reports are other qualitative analysis reports are generated and analyzed.
8. **Implementation**: The system is deployed into a production environment.
9. **Day to day Operation and Maintenance**: This is the post deployment phase where the operational life of the system is reviewed and maintained if needed.
10. **Disposition**: This refers to the end of system activities related to the product.

So basically the model of the System Development Life Cycle has the following major components:

1. Requirement Analysis:
2. Design:
3. Implementation
4. Testing:
5. Evolution:

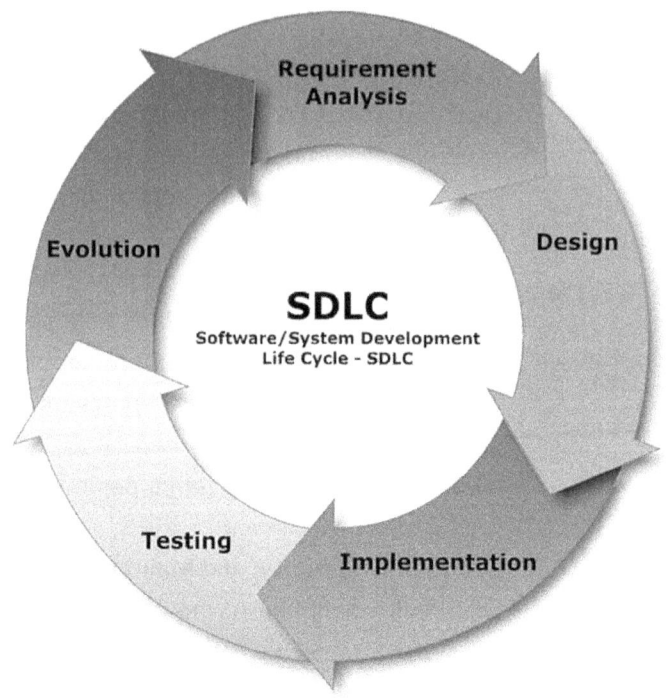

Fig: 5.a

Waterfall model

In this process the developers follow the different phases described in the previous section in order.

In this model once one phase is finished, it proceeds to the next one. Reviews may occur before moving to the next phase which allows for the possibility of changes.

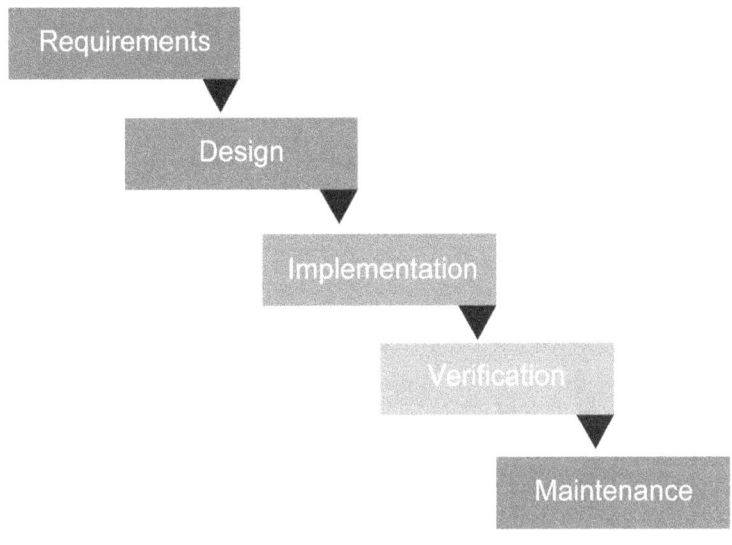

Fig: 5.b

Reviews may also be employed to ensure that the phase is indeed complete; the phase completion criteria are often referred to as a "gate" that the project must pass through to move to the next phase. Waterfall discourages revisiting and revising any prior phase once it's complete.

Spiral model

In this model deliberate iterative risk analysis, particularly suited to large-scale complex systems happens at a predefined frequency. It emphasizes risk analysis, and thereby requires customers to accept this analysis and act on it. So the developers typically spend more to fix the issues and are therefore often used for large-scale internal software development.

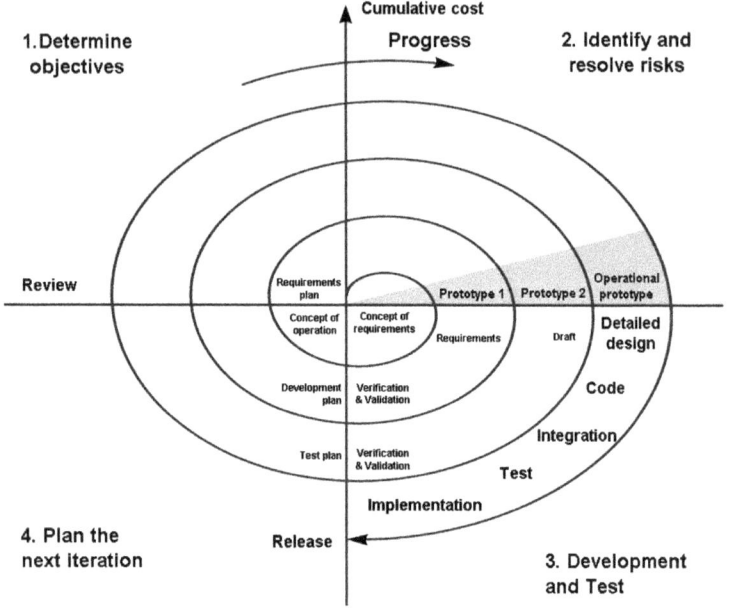

Fig: 5.c

The Spiral is visualized as a process passing through some number of iterations, with the four quadrant diagram representative of the following activities:

1. Formulate plans to: identify software targets, implement the program, clarify the project development restrictions
2. Risk analysis: an analytical assessment of selected programs, to consider how to identify and eliminate risk
3. Implementation of the project: the implementation of software development and verification

Because of frequent risk analysis and more effort spent by the developer to analyze the risks accurately, the cost factor goes up in the project.

Iterative Development Model

This method helps to develop a system through repeated cycles and in smaller chunks at a time, allowing software developers to take advantage of what was learned during development of earlier parts or versions of the system.

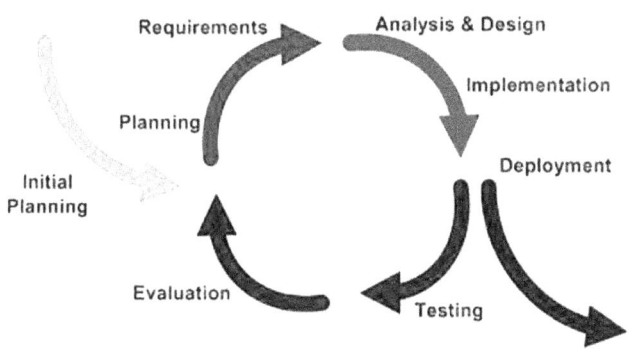

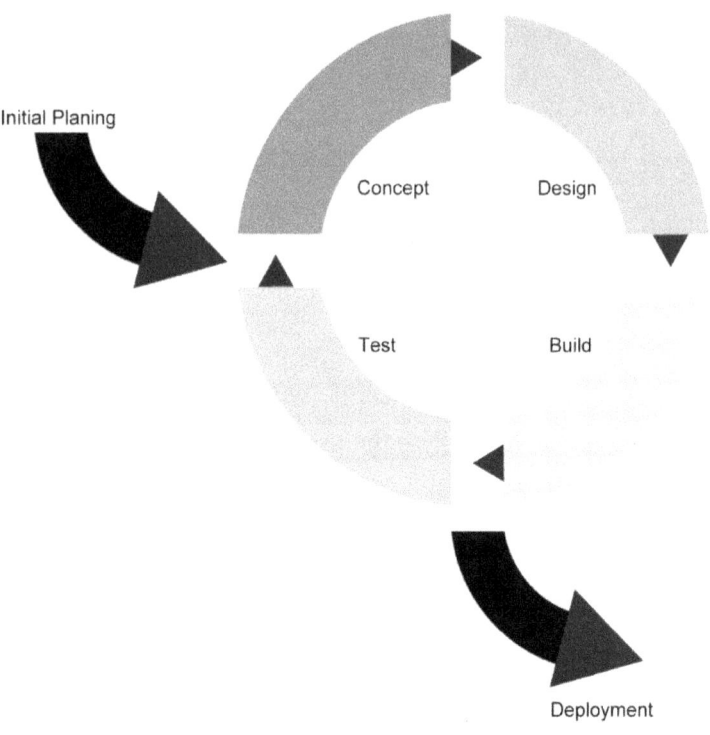

Initial Planing

Concept

Design

Test

Build

Deployment

Fig: 5.d

Incremental development divides the system functionality into increments (portions). In each increment, a portion of functionality is delivered through cross-discipline work, from the requirements to the deployment. The unified process groups increments/iterations into phases: inception, elaboration, construction, and transition. It identifies scope, functional and non-functional requirements and risks at a high level which can be estimated.

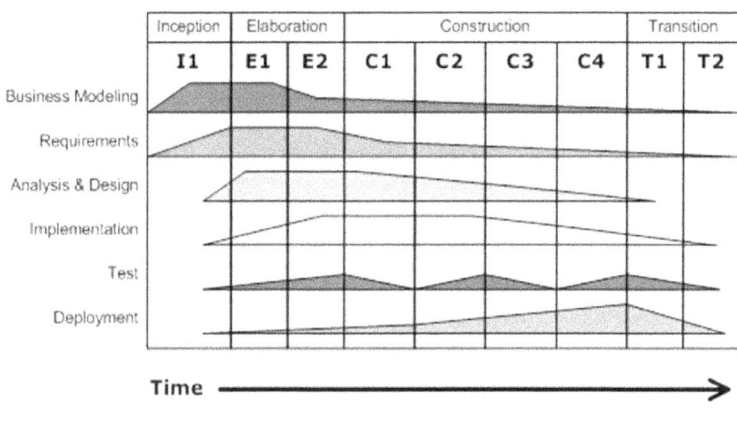

Iterative Development

Business value is delivered incrementally in time-boxed cross-discipline iterations.

	Inception	Elaboration		Construction				Transition	
	I1	E1	E2	C1	C2	C3	C4	T1	T2
Business Modeling									
Requirements									
Analysis & Design									
Implementation									
Test									
Deployment									

Time ⟶

Fig: 5.e

Applying this model to multidisciplinary complex project with large volume can come with a risk as inability in the developers part to uncover important issues early before problems can spoil the project.

Agile development

This is perhaps today's the most widely used SDLC model. It uses iterative development as a basis ,but uses people-centric viewpoint through user feedbacks rather than planning as the primary control mechanism. The feedback is driven by regular tests and releases of the evolving software.

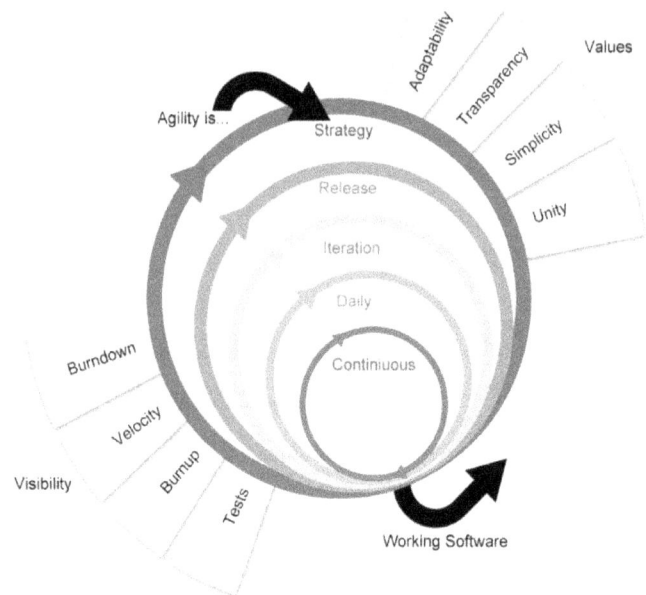

Fig: 5.f

There are many variations of agile processes:

1. Agile Data (AD)
2. Agile Microsoft Solutions Framework (MSF)
3. Agile Modeling (AM)
4. Agile Unified Process (AUP)
5. Dynamic System Development Method (DSDM)
6. Extreme Programming (XP)
7. Feature Driven Development (FDD)
8. Scrum
9. Usage-Centered Design (UCD)

Challenges in UX Integration to Different SDLC Models

However there are several challenges in integrating UX design and related activities into a typical agile software development life cycle process. The most common problem is typically " finding a balance between up-front interaction design and integrating interaction design with iterative coding with the aim of delivering working software instead of early design concepts". This happens mostly because typical pure SDLC approaches primarily aim at "efficient coding tactics together with project management and team organization instead of usability engineering".

However even though there are many challenges in integrating UX design with agile practices, some researchers see "agile software development practices" as enablers for bringing UX design closer to software engineering and enhancing interaction between these two disciplines.

In Current State of Agile User-Centered Design: A Survey. HCI and Usability for e-Inclusion. Lecture Notes in Computer Science, (by Hussain, Z., Slany, W., and Holzinger, 2009), a survey of the integration of usability and UCD practices with agile methods reported that "the majority of the respondents found that usability and user-centered design practices had brought added value through improvements in the usability and quality of the end product. Development teams often report that they are better able to respond to the needs of the customer with agile methods, and their measured or perceived productivity has been reported to be better than development teams using traditional methods".

Usability Designing Process

While exploring through available scopes in different SDLC models, we can notice a set of common activities which can be combined together to form a generic Usability Designing Process. Basically there are four phases of the whole set of activities which can be tagged separately as follows:

1. **Plan**: This phase involves the following major activities:
    i. Developing a Plan
    ii. Assembling a project team
    iii. Kicking of the project
2. **Analyze**: This can have the following activities:
    i. Evaluation of scope
    ii. Evaluation of existing product (if enhancements are planned)
    iii. User Research
    iv. Task Analysis
    v. Persona Development
    vi. Scenarios evaluation and prescription
    vii. Define measurable goals
3. **Design**: This is phase where some out puts related to the actual product are generated:
    i. Product/ Site requirement analysis
    ii. Conducting a content inventory
    iii. Performing card sorting

iv. Information Architecture (IA) formation
v. Writing for web
vi. Parallel design
vii. Wire framing and prototyping
viii. Usability Consultations to programming team

4. **Test and refinement**: This is a phase that can be applied to multiple phases of SDLC as it is mostly about usability testing :

i. Usability Testing
ii. Heuristic Evaluations
iii. Implementation and retest

Note that the above phases are not the phases we discussed regarding different SDLC processes. Rather these are the phases that can paired with different phases of SDLC processes depending on the SDLC model used. Typically Usability Designing Process is led by the "usability designer" with the team of field study specialists, user research specialists, usability evaluation specialists, prototype developers, interface designers and graphic designers along with other specialists from related disciplines that varies from project to project.

How Usability Design Process Fits in Different Phases of SDLC?

The different blocks of usability design process can be mapped to different process blocks of any SDLC model. The following diagram shows a generic model for typical process blocks for the typical iterative SDLC model.

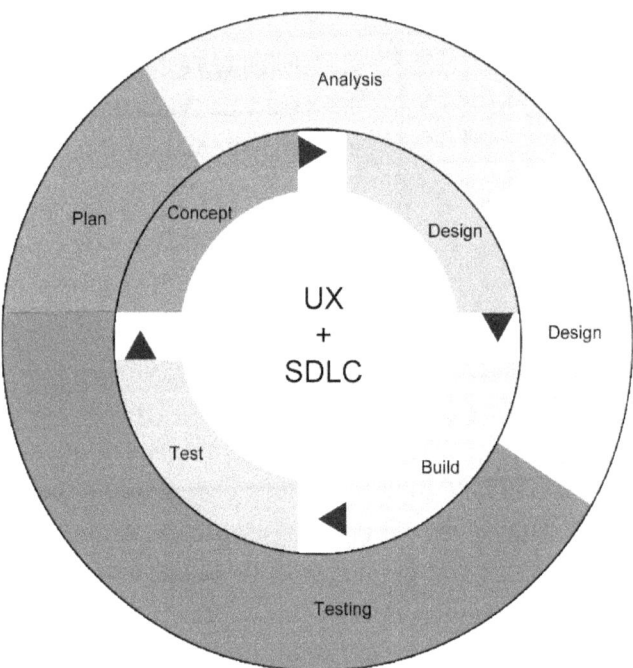

A generic model representing different phases of Usability Process blocks being implemented along with a generic SDLC process blocks. The outer circle reprents UX process blocks where as the inner circle represents SDLC process blocks.

Fig: 5.g

One thing to keep in mind is that process is not a complete product development process as it does not out put the final product at the end of the process cycle. Rather Usability process supplements to any software development life cycle at various stages.

Plan or Ideation and Requirement Analysis Phase: In this phase while the technical feasibility is being evaluated, UCD contributors can assist product management by conducting user research with people in the target market to evaluate target user goals, tasks, and workflow. Requirement analysis phase is also aided by the UCD experts in careful review of the gathered data and preparing metrics that can be used in the development phase.

Design Phase: UCD contributors who are skilled at interaction design, visual design, and information design create mockups or prototypes of portions of the system, and contributors who are trained in usability evaluation assess the designs by subjecting them to usability testing.

Build or Development Phase: UCD contributors are usually called upon in a consultative or interpretive role, meeting with the developers responsible for actual implementation of the product, and providing guidance for underspecified areas of the product. In this phase the UCD contributor's role is to remain the consistent user advocate throughout the project. When negotiations must happen during design and development of a feature, the UCD contributor reminds the team of the design persona (the "design target", or user group at which the feature is aimed), helps the product manager identify and weigh the risks of leaving off certain areas of functionality, etc.

Integration and Testing/ Verification Phase:UCD experts help in benchmarking usability tests popularly known as "summative evaluations" that evaluates performance of the system /product developed on several grounds. The metrics of this test is typically based on the "error rate for users as they use the system", the "time it takes to attain proficiency performing a task", and the "time it takes to perform a task once proficiency has been attained".

How UX Fits in Different Models of SDLC?

The pre-agile era saw many attempts of UX getting fitted into the waterfall and it's derivative models of SDLCs. Such attempts by the many developers were natural outcome of the post-projects disasters, where 'design' was never the personality of "software product engineering" and the lack of usability doomed the products even after the initial set of requirements check list was fulfilled.

More demands for Graphic User Interfaces (GUI) in software application (due to GUI's power to offer better visibility and power to the end users) tempted the developers to follow emerging UX practices which included a task to add "design phase" to existing SDLCs. Waterfall model was good enough to accommodate a notion of design in its phases and become popular despite its limitations (which later pave paths for Agile era). Most of the design approaches and techniques created during this era were having mostly a goal "to eliminate any deviation during the development process, by telling the developers exactly what we expect of them".

Let's see how different models of SDLC accommodated UX differently in the following:

Waterfall Model: Historically the waterfall model of SDLC can use the UCD components in its engineering process and the product to translate the "set of requirements into something beautiful". It is relatively simpler and easy to spot the to spot the areas within different phases where UX can be easily fits in as each phases are clearly demarked for different activities.

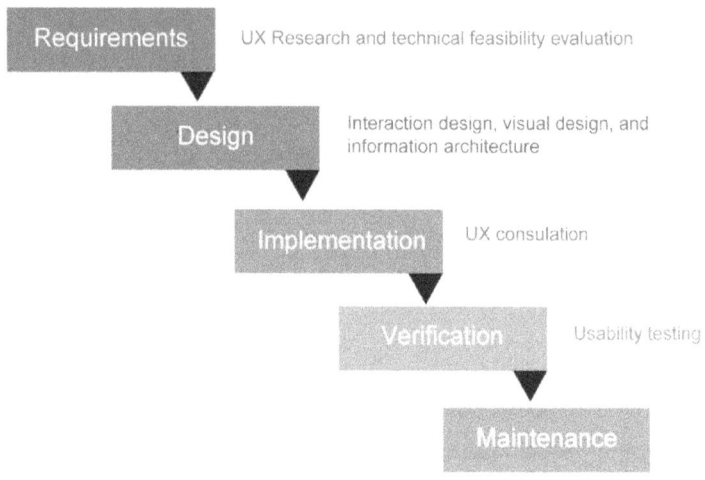

Fig: 5.h

Spiral Model: In Spiral model, the UCD design can work across different quadrants of activities. The first quadrant where the objectives are determined, the usability and user research can happen as this is where requirements are planned. In the second quadrant the activities involving risk identification can best use UX activities involving IA and prototyping. The third quadrant of development and testing can utilize consultation and usability testing. The final fourth quadrant of activities can be used for feasibility evaluation and setting up usability metrics and bench marking for the next release.

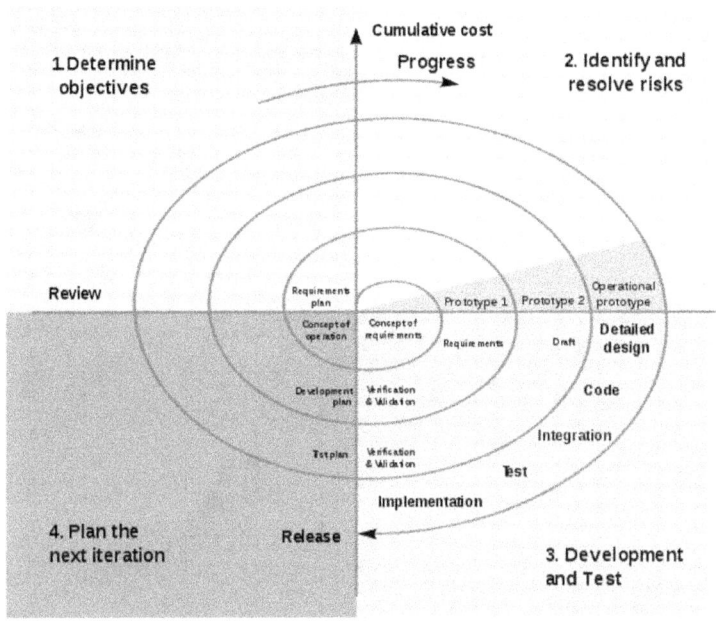

UX Research

Interaction design, information architecture and prototyping

UX consulation and Usability Testing

Technical feasibility evaluation and usability metrics

Fig: 5.i

Iterative Model: In this modeleach iteration cycle can bedivided into different activities phases to incorporate UCD methodologies for UX integration. Each iteration activities block that are mostly split across concept, design, build and test phases can be used for different UCD activities .

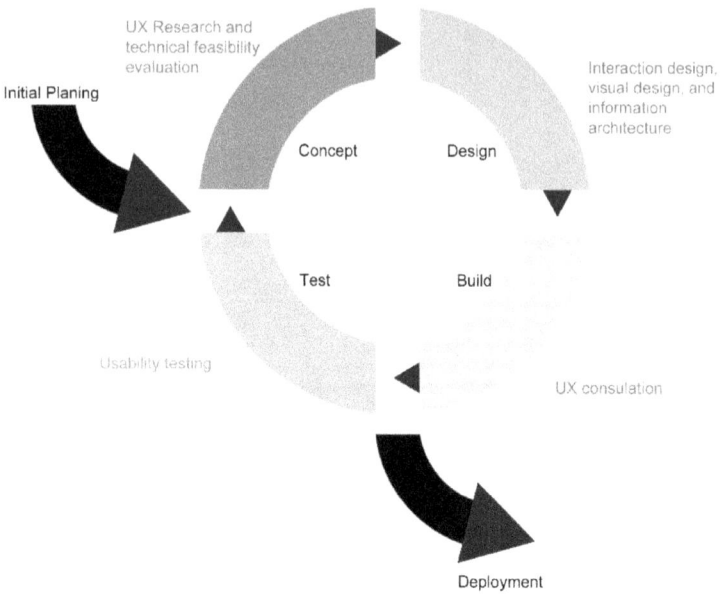

Fig: 5.j

Agile Model: This is the most popular and successful SDLC model as of today as it allows better scope in providing continuous and iterative refinement to the product.

Historically when developers out of their frustrations with waterfall model turned to the growing Agile Movement to regain their control over the process, they found that "like its ancestors, Agile also didn't take UX into account. Several of the Agile methods, such as Scrum and XP, recommended users sitting with the team during the development process, but that isn't the same as design. Everyone who figured out how to get what they wanted from plugging UX into a phased waterfall approach was now struggling to work inside the Agile methods. The Agile principles, that focus more on communication and less on contracts, didn't fit the status quo UX processes".So efforts were made again to implement UX into Agile methods just like the way it was implemented into waterfall model . But it was not easy as , in waterfall model there are 2 things which helped implemented UX :

1. The objectives of the project stays same from kickoff to the point where the finished product is launched.
2. The designers created the set of design specifications as a contract which the developers had to implement into the final product.

And above two cannot be expected from Agile model as it is based on iterations and gradual exploration of what is best fit for the final product . On ejust simply cannot predict the final design from the start of the project. So many attempts were made to get the best agile SDLC practices that can incorporate the UX , before "Lean UX" was born.

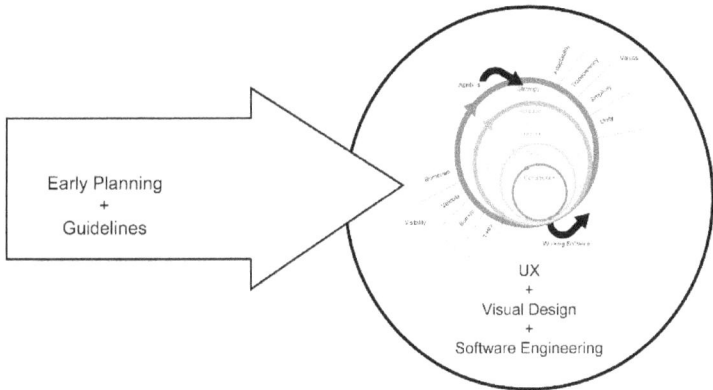

Fig: 5.k

As the above figure shows the documentation and guidelines are stripped to their bare minimum components, providing the minimum amount of information necessary to get started on implementation. Also Long detailed design cycles are discarded in favor of very short, iterative, low-fidelity cycles, with feedback coming from all members of the implementation team early and often.

Challenges with Agile model of SDLC to implement UX

There are several challenges in implementing UX in Agile model effectively and these challenges include:

1. **Different approaches.** Usability methodologies are centered on the user and holistic view of the user needs whereas agile methodologies take a broader view and focus on the stakeholder. Agile methods primarily focus on delivering working software early.

2. **Different goals.** Software engineers focus on the technical design, implementation the system where as UX practitioners focus on "developing systems so end-users can use them effectively".

3. **Organizational challenges.** The agile methodologies focus on strategy where teams are self-organizing whereas UX focuses on a centralized UX groups within some organizations so that the needed practices, tools, and standards can be provided. .

4. **UX practitioners struggle to be heard.** Many UX practitioners often complain that the results of their work are not considered in the design decisions and even if it is heard, there seems to be focus more on engineering decisions over the usability decisions.

Lean UX and Agile Model

Many of UX practitioners see "Lean UX" as the answers to the challenge we see in implementing UX into an Agile SDLC where it uses "taking the best parts of our current UX practices and redesigning them specifically for use in an agile world". Lean UX, is about reducing waste in a process by removing it from the value chain of the usability process..

The proactive measures for border engagement in Agile model has paved path to a new and more practical implementation of UX discipline and methods called "Lean UX" Lean UX once blended with any exiting Agile SDLC, helps to "create a more productive team and a more collaborative process".

The basic principles Lean UX uses to provide positive refinements to SDLC are through the following 3 foundation stones for it:

1. **Design Thinking:**This foundation upholds the concept that "every aspect of a business can be approached with design methods" and gives "designers permission and precedent to work beyond their typical boundaries".
2. **Agile Software Development**: Core values of Agile are the key to Lean UX.
3. **Lean Startup method:**Lean Startup uses a feedback loop called "build-measure-learn" to minimize project risk and gets teams building quickly and learning quickly

We will explore more about "Lean UX" in in coming chapters.

Agile in Usability Design: Without Reference to SDLC

This chapter is not about how usability process is related to agile software development life cycle, rather in this chapter we are going to see how iterative and agile methodology can be used into the Usability design process. Both are two different concepts at least have two different contexts.

Usability Designing Process

In last chapter we briefly explored the basic steps of "Usability Design process".

1. **Plan**: This phase involves the following major activities:
    i. Developing a Plan
    ii. Assembling a project team
    iii. Kicking of the project
2. **Analyze**: This can have the following activities:
    i. Evaluation of scope
    ii. Evaluation of existing product (if enhancements are planned)
    iii. User Research
    iv. Task Analysis
    v. Persona Development
    vi. Scenarios evaluation and prescription
    vii. Define measurable goals
3. **Design**: This is phase where some out puts related to the actual product are generated:
    i. Product/ Site requirement analysis
    ii. Conducting a content inventory
    iii. Performing card sorting

iv.   Information Architecture (IA) formation
v.   Writing for web
vi.   Parallel design
vii.   Wire framing and prototyping

4. **Test and refinement**: This is a phase that can be applied to multiple phases of SDLC as it is mostly about usability testing :
   i.   Usability Testing
   ii.   Heuristic Evaluations
   iii.   Implementation and retest

Lean UX:
Another Agile UX?

"Lean UX" is seen as the answers to the challenge we see in implementing UX into an Agile SDLC. As discussed in the earlier chapters, Lean UX principles use "taking the best parts of our current UX practices and redesigning them specifically for use in an agile world".

Lean UX solved many issues were even there with the practices and usability process used in waterfall and different derivative iterative models. In Lean UX practices, there is no more requirements for the "objectives will stay the same and that you can design for a single solution throughout the project". Also there no more need for the designers to create bulky guidelines and documentations for the developers to be used as a contract. Rather the Lean UX practice upholding agile approach, only focuses on each independent design iteration as a "hypothesis" which has to be validated from a "customer perspective" and from a "business perspective" . The beauty here is because of the Agile process being followed, fast iterations can happen to quickly test hypotheses and get to a great design in the end of the project.

In context of a real world situation the Lean UX is something like:

> "The traditional paper work is discarded, while the focus is turned to making sketches of the idea. Then the sketch is presented and discussed with the team. The initial prototype effort is very small comparing to detailed documents, so it's easy to make changes. After it's agreed internally, rough prototype is made and

tested with users. The learning from users help refine the idea and iteration starts over again"

And this makes case for "collaboration with the entire team" as it becomes critical to the success of the product and the whole project.

The Beauty of Lean UX: Everything is familiar

No practice used in Lean UX is something new. Rather it is "built from well-understood UX practices". Many of the techniques used over the time in various UX process and have the practical usability even today, have been packaged properly in Lean UX.

Lean UX is not Same as Agile UX

Lean UX is a totally different term than Agile UX.Lean UX details methods and their practical application in dynamic environment of a Lean StartupIt is the converging point for product development and business, through constant measurement and so called "learning loops" (build – measure – learn).
Agile UX defines update of Agile Software Methodology with UX Design methods. It aims to unify developers and designers in the Agile process of product development.

However Lean UX uses Agile UX methodologies, tools to coordinate their software development.

Foundation Stones of Lean UX

The key ingredients of Lean UX which act as foundation stones for it are:

1. **Design Thinking:**This foundation upholds the concept that "every aspect of a business can be approached with design methods" and gives "designers permission and precedent to work beyond their typical boundaries".
2. **Agile Software Development**: Core values of Agile are the key to Lean UX. It forces on four major principlesof Agile development to product design:
    i. **Individuals and interactions over processes and tools**: This principle louds the concept of exchange of ideas freely and frequently in a team. "The constraints of current processes and production tools are eschewed in favor of conversation with colleagues"
    ii. **Working Software over Documentation:**This focuses on bringing out solution quickly so that it can be "accessed for market and viability".
    iii. **Customer Collaboration over Contract Negotiation:**Collaboration with teammates and customers builds consensus behind decisions which

results in faster iterations and lessens heavy documentations.

iv. **Responding to change over following a plan**: "The assumption in Lean UX is that the initial product designs will be wrong, so the goal should be to find out what's wrong with them as soon as possible" and this helps in finding the right direction quickly.

3. **Lean Startup method:** Lean Startup uses a feedback loop called "build-measure-learn" to minimize project risk and gets teams building quickly and learning quickly. Typically startups are "free to build products in a manner which differentiate on quality". Also startups can focus on "intrinsic value and usefulness of the product, rather than on a long list of mostly irrelevant features". Startups have a distinguishable feature of reducing long product cycles into smaller, shorter chunks and validating these iterations with people that will use the products, actually opens the gate for important information needed to avoid expensive development cycles that come withsome kind of risk. "The secret sauce of lean startup people is that they advocate for user experience research and design as one of the primary solutions to their business problems, and they do it using plain language".

Lean Startup method: The concept of "Build-Measure-Learn"

The fundamental activity of a startup is to "turn ideas into products" at the first step. Next it has the aim to measure how customers respond and then to learn whether to pivot or persevere. So basically a startup's success depends on this feedback loop. To be successful a startup has to accelerate that feedback loop. The feedback loop being employed here includes three primary activities: build the product, measure data and learn new ideas.

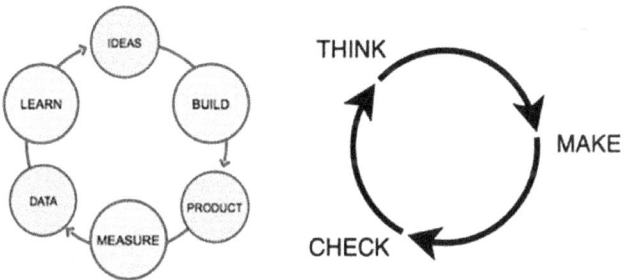

Fig: 7.a

However the Lean Startup method employed in Lean UX , is slightly different – it's basically about "Think-Make-Check". The difference lies in the fact that in latter case the "feedback loop incorporates your own thoughts as a designer, not just ideas learned through measurement".

Minimum Viable Product (MVP) – Prototyping at it's best in Lean Startup Method

Minimum viable product is a version of the product that enables a full turn of the Build-Measure-Learn loop with a minimum amount of effort and the least amount of development time. The MVP is, in fact, an early prototype that serves as a tool to learn and test the team's assumptions.

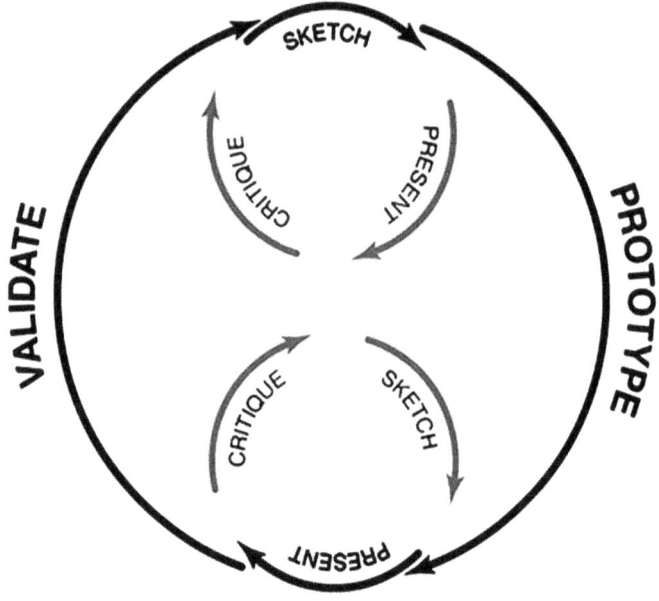

Fig: 7.b

Lean UX is where prototyping is best promoted, focusing the prototype on major components of the experience. Once created, it will be immediately testable by any and all users to start the feedback loop. In most of the case the fidelity of the prototype is not a road block, rather the only mission is to get a quick prototype that can be tested quickly. However where the need for better visualization has priority, the best practices and tools are used to develop high fidelity prototypes in shortest time possible.

Principles of Lean UX

There are some guiding principles behind Lean UX which can be used to make sure the methodologies used in a Lean process is on track.

1. **Cross-Functional Teams:** Specialists from various disciplines come together to form a cross functional team to create the product. Such a team typically consists of Software engineering, product management, interaction design, visual design, content strategy, marketing, and quality assurance (QA).
2. **Small, Dedicated, Collocated:** Keep your teams small— no more than 10 total core people as keeping small team has the benefit of small teams comes down to three words: communication,focus, and camaraderie. It is easier to manage smaller team as keeping track of status report , change management and learning.
3. **Progress = Outcomes, Not Output:** The focus should be on business goals which are typically are the "outcomes", rather than the output product/system or service.
4. **Problem-Focused Teams:**"A problem-focused team is one that has been assigned a business problem to solve, as opposed to a set of features to implement".
5. **Removing Waste:** This is one of the key ingredients of Lean UX which is focused on "removal of anything that doesn't lead to the ultimate goal" so that the team resource can be utilized properly.

6. **Small Batch Size:** Lean UX focuses on "notion to keep inventory low and quality high".

7. **Continuous Discovery:** "Regular customer conversations provide frequent opportunities for validating new product ideas"

8. **GOOB: The New User-Centricity: GOOB stands for** "getting out of the building" -- meeting-room debates about user needs won't be settled conclusively within your office. Instead, the answers lie out in the marketplace, outside of your building.

9. **Shared Understanding:** The more a team collectively understands what it's doing and why, the less it has to depend on secondhand reports and detailed documents to continue its work.

10. **Anti-Pattern: Rock-stars, Gurus, and Ninjas:**Team cohesion breaks down when you add individuals with large egos who are determined to stand out and be stars. So more efforts should on team collaboration.

11. **Externalizing Your Work:**"Externalizing gets ideas out of teammates' heads and on to the wall, allowing everyone to see where the team stands".

12. **Making over Analysis:** "There is more value in creating the first version of an idea than spending half a day debating its merits in a conference room".

13. **Learning over Growth:** "Lean UX favors a focus on learning first and scaling second".

14. **Permission to Fail:** "Lean UX teams need to experiment with ideas. Most of these ideas will fail.The team must be safe to fail if they are to be successful".

15. **Getting Out of the Deliverables Business:** "The team's focus should be on learning which features have the biggest impact on the their customers. The artifacts the team uses to gain that knowledge are irrelevant."

Bibliography

Cooper, Alan, Reimann, Robert and Cronin, Dave. 2007. *About Face 3:The Essentials of Interaction Design* Wiley Publishing. Inc.

Garrett, Jesse James. 2002. *The Elements of User Experience*. New Riders Press

Lowdermilk, Travis. 2013, *User Centered Design*, O'Reilly

Buxton, Bill. 2007, *Sketching User Experiences- getting the design right and the right design*, Diane Cerra

Julie A. Jacko (Ed.), 2007, *Human-Computer Interaction – Interaction Design and Usability*, Springer

Beyer, Hugh, 2010, *User-Centered Agile Methods*, Morgan & Claypool

Gothelf, Jeff and Seiden, Josh. 2013, *Lean UX- Applying Lean Principles to Improve User Experience*, O'Reilly

About the Author

Samir Dash (b. 1982-) has more than 10 years of experience in information design solutions for both software & IT services as well as Fortune 100 product setups. His experience includes strategic road-map and planning for e-learning, social and mobile software products & services; defining product and UX strategy for mobile, SaaS and cloud based eco-systems; RIA and front end development using open technologies like HTML5, Java Script. Samir has been recognized with awards like Manthan Award, World Summit Award and M-Billionth, for some of his experimental mobility projects.

Samir can be contacted via his Linked profile
http://in.linkedin.com/in/mobilewish/

www.ingramcontent.com/pod-product-compliance
Lightning Source LLC
Chambersburg PA
CBHW051220170526
45166CB00005B/1976